Continents of the World for Kids

Geography for Kids

World Continents

By

Nishi Singh

Copyright:

Copyright © 2016 Nishi Singh, PhD. All Rights Reserved.

http://www.nscientific.com

No portion of this book may be modified, reproduced or distributed mechanically, electronically, or by any other means including photocopying without the written permission of the author.

Disclaimer:

Whilst every care is taken to ensure that the information in this book is as up-to-date and accurate as possible, no responsibility can be taken by the author for any errors or omissions contained herein. While most of the images used in this book belong to the author, some are in the public domain compiled from various image repositories. Research for the book has been done using reliable sources and the author's own personal experience. In the event where any fact or material is incorrect or used without proper permission, please contact us so that the oversight can be corrected.

About the author

Nishi Singh graduated from the University of West London (Brunel University) with a BSc. (Hons) in Applied Biochemistry and an MSc. in Molecular Pharmacology from the University of Manchester and immediately proceeded to do a PhD. at Cardiff University. After completion of my PhD., he was awarded a postdoctoral position funded by the British Heart Foundation to continue cell signalling research.

Other books by the author

http://www.amazon.com/author/nishisingh

Cells For Kids (Science Book For Children)

Fart Science: All About Fart: Farting, Passing Gas And Flatulence

HeLa Cells of Henrietta Lacks

Table of Continents

About the author 3
Chapter 1: Introduction 5
Chapter 2: Asia 8
Chapter 3: Africa 14
Chapter 4: North America 20
Chapter 5: South America 26
Chapter 6: Europe 32
Chapter 7: Australia 38
Chapter 8: Antarctica 43
Chapter 9: Conclusion 48
References and further reading 51

Chapter 1: Introduction

The world may seem like a big place, but it is actually very small. The planet Earth that we reside on is made up of oceans and continents. A continent is basically a giant landmass that is comprised of different countries. Most people think of continents as being separated by huge bodies of water, like oceans. When you look at a map you can easily see that ocean water separates the majority of continents from one another. But despite this popular belief, there is no strict criterion that separates continents from each other. The different continents are identified from one another by convention, which are regional standards that are determined by people. This convention has determined there to be seven continents throughout the world. These continents are Africa, Asia, North America, Antarctica, South America, Australia and Europe. People think about these continents in different ways. For example, some people think about continents based on the typical climate in the region. If look at Antarctica people often think about the freezing cold temperatures and the snowy conditions on the ground. If you look at South America people think about the hotter temperatures because it is a continent near the equator where temperatures are very warm. Then there are continents like North America and Europe which are more popular for their cultural aspects than their temperatures.

All the continents in the world have coasts around them that lie in front of an ocean. The only two continents that are isolated from all the other continents are Australia and Antarctica. The other continents are at least joined by some small amount of discrete landmass. For example, the African and Asian continents are joined together by a 75 mile strip of land called the Isthmus of Suez. The South American and North American continents are joined together by the Isthmus of Panama, which is a narrow strip of land that happens to make up the country of Panama. This country is really the divider of the two continents, even though it is conventionally considered to be part of North America. As you can see, there are so many great things to learn about these continents and that is what this book is all about. You will learn everything you've ever wanted to know about these seven continents and how they are important to this world. You may never get to visit all seven continents, but being knowledgeable in them will help you understand the geology of the Earth and how civilization came to be what it is today. After all, the position of the continents on the Earth has helped shape human cultures, agriculture, religions and our way of life. By learning about the continents, you will be able to share this knowledge with your friends and get them interested in studying the world's continents as well. Then perhaps when you grow up you will want to travel to these regions of the world and see them for yourself because they are truly fascinating when experienced in real life. But for now, it is time to learn about them.

Chapter 2: Asia

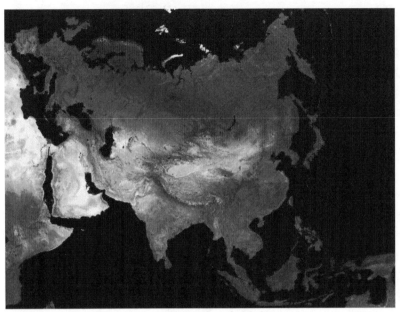

The continent of Asia

When people hear the word "Asia" they typically think of popular Asian countries like China or Japan. But what people forget is there are 48 countries that make up the Asian continent, and many of them are very diverse from one another. Asia is also the largest continent in the entire world and has more people living there than on any other continent. Statistics show there are currently 4.4 billion people living in Asia. Even though Asia's landmass only makes up about 9% of the total surface area of the Earth, it does make up 30% of the world's total land mass. This is the continent where the population of human beings has increased the most. This is why some Asian countries, like China, have laws in place that limit the number of children each family can have because they need to control their overpopulation problem.

The author at the Taj Mahal in the city of Agra in India (Asia)

The interesting thing about Asia is there is not much water separating it from the European continent. The Asian boundaries have no clear geological features that separate them apart. The only features available are the ones that humans have decided upon which separate the continents. The main boundaries that make Asia an eastern continent are the Ural River, Ural Mountains and the Suez Canal. Then to the south you have the

Black Sea, Caucasus Mountains and the Caspian Sea. Now what makes Asia even more interesting is that it has the most cultural, environmental, historical and economical differences throughout its landmass than any other continent. For example, did you know that India is considered to be part of Asia? When you think of an "Asian" person you don't usually think about somebody who is from India. But technically an Indian is also an Asian because they all come from the same continent. India is in southern Asia, so you might refer to them as southern Asians if you want to get more specific when describing them. As for Russia, this country is why the term "Eurasia" exists. The majority of Russia's landmass falls within the Asian continent, but it also has land that falls within the boundaries of the European continent. In fact, the Russian capital "Moscow" lies within the eastern region of the Europe. So when people think about Eurasia they are usually thinking about Russia and how it lies within both continents.

Historically, the ancient Greeks were responsible for the border that separates Asia and Europe. They used the Aegean Sea, Sea of Marmara, the Dardanelles and the Kerch Strait to separate the two continents. As for Africa and Asia, Greek geographers determined the Red Sea was the true boundary between those two landmasses in the 15th century. Before this time period, they used the Nile River as the boundary line. So as you can see, there is no truly set boundary line for any continent. The boundary is whatever people want it to be through the help of geological features. The ancient Greek civilization was one of the first civilizations to establish these kinds of boundary lines and they did it starting with Asia and Europe. In fact, the word "Asia" comes from the ancient Greeks. But the word's original meaning seems to be unknown because the word can be seen in different languages and has different meanings in those languages. During the days of the Roman Empire, the ancient Romans named one of their provinces "Asia." This province is now where modern day Turkey is located. But it was the Greeks that first used the name to describe the continent and it has been called that ever since.

Ural Mountains that is part of the boundary between the continents of Europe and Asia

Asia is divided into four regions; East Asia, Middle East, South Asia and Southeast Asia. They are all linked by the huge landmass in the Central Asian region called steppes. This is a region inhabited mostly by wild horses because there is plenty of room for them to run around and plenty of grass for them to eat. Through the Central Asian steppes, you can reach all four regions of Asia. Everything surrounding the steppes is either desert or mountains. The climate throughout these regions is also quite diverse. If you look at the Middle Eastern region, this is a desert climate that experiences mostly hot temperatures. As you go towards

Southeastern Asia you will find countries like India that are dry and hot. But then as you go towards East Asia you will find a much colder climate with lots of snow and icy mountains. Most of Russia lies within East Asia, which is why the country is known for being very cold and snowy. The diversity of the continent along with its huge size makes it very unique in the world. Millions of tourists travel throughout Asia each year because of its diversity and historical sites. With the second biggest economy, Asia helps stimulate the economy of North America and every other major country's economy on Earth.

Chapter 3: Africa

Map of Africa

The African continent is said to be the oldest continent in the world. This is the continent where all life on Earth began, particularly human life. Scientists have already figured out that human life started in Africa and then gradually spread to other continents because humans migrated to them. But central eastern Africa seems to be accepted by scientists as the specific location in Africa where human life began.

Africa currently has the second biggest population out of any other continent in the world. It is also the second largest continent in the world. A recent census found there are currently 1.1 billion people living on the African continent. Africa covers about 6% of the total surface area of the Earth and about 20.4% of all its landmass. The boundary lines of Africa can all be found in the northern section of the continent. This would be the Mediterranean Sea, Red Sea and the Suez Canal. In the south you just have large bodies of water, like the Indian Ocean in the southeastern region and the Atlantic Ocean in the western region.

Africa is very easy to identify when you look at it from a world map. It is a huge landmass in the center of the globe that covers about 11.6 million square miles. The equator runs right through the African continent. As you probably know, any area close to the equator is going to have very hot temperatures because these are the areas which get hit by the sun the most. If you look at Antarctica, for example, it is on the very bottom of the Earth and as far away from the equator as you can get.

That is why it is so cold and why Africa is so hot. In fact, you've probably heard about Africa having all kinds of wild animals and exotic insects living there. The reason they inhabit this particular landmass more than any other landmass is because of the hot climate. If it were colder then these life forms wouldn't be able to survive. Warm weather is the reason why life began on Earth in the first place, which explains why all human life started in Africa.

The Egyptian pyramids in North Africa

Many people refer to Africa like it is all one big country, but the truth is it has 54 separate countries inside of it. These countries are typically referred to as

16

sovereign states, which is likely the reason why people confuse the continent for being an entire country. But the term "sovereign state" is another way of describing a country because they both have one centralized government that regulates a specific geographic area. The African continent may be all one big landmass, but has been broken up into 54 countries and 9 territories throughout the land. The largest country in Africa is Algeria, but the country with the biggest population there is Nigeria. However, all throughout Africa you will find a great variety of ethnicities, languages and cultures. You probably know that Africa has a large population that is of the black race, but over the many centuries there has been colonization and immigration on the continent. This brought people of different cultures and races from other continents to Africa. Along with them, they brought their languages and religions as well.

During the late 1800s, many settlers from the European continent colonized the majority of the African continent. This means they went to Africa and enforced their own set of rules and laws over the people that were already there. But then in the 20th century, there was a decolonization process which allowed the native African people to take back their continent and establish their own rule of law. One famous example of this is with South Africa, which is the southernmost country of the continent. This country was ruled by Europeans for hundreds of years and hand white presidents ruling the land. Then in 1994, the first black president took office named Nelson Mandela. Another interesting aspect about Africa is the country of Egypt in the north. You probably know Egypt for being the place with those big pyramids. These pyramids are thousands of years old and were created as burial grounds for the Egyptian rulers, called pharaohs, during this time period. Even to this day, these historical sites still exist and are visited by millions of tourists from around the world each year.

As far as the African economy goes, it is certainly one of the poorer continents in the world. In fact, Africa has the most countries with people in poverty and starvation than any other continent. Their technological and educational growth still has to catch up with Europe, Asia and North America. But Africa does have a big agricultural industry because of the rich soil that produces high quality crops. This allows them to stay somewhat competitive in the international marketplace as far as produce is concerned. The rest of Africa's economy thrives on tourism. Do not be fooled by the common misconception that all of Africa is made up of poor and desolate areas. There are actually many cities throughout the countries of Africa that are as big as Chicago. Perhaps in another 100 years, many of its more populous countries will catch up to the rest of the world in technological innovation and economic growth.

Chapter 4: North America

Continent of North America

The continent of North America lies completely within the Northern Hemisphere of the Earth and mostly in the Western Hemisphere. Some people call it a subcontinent of the Americas, since South America is in the south and North America is in the north. But each landmass has been determined to be separate continents. The boundary line between the two continents is the country of Panama. Even though it is technically considered part of North America, everything south of Panama is the South American continent. You also have southern boundaries that include the Pacific Ocean and the Caribbean Sea. The other boundaries in North America are the Atlantic Ocean in the east and the Arctic Ocean in the north.

Statue of Liberty on Liberty Island in New York City (United States)

North America is the third largest continent in the entire world. It is made up of 23 different countries and contains about 565 million people. The three biggest and most popular countries in the continent are Canada, the United States and Mexico. The United States is by far the richest country in North America. In fact it is one of the richest countries in the entire world. The United States is an interesting country because it is split between the colder and hotter climates. The majority of the country lies in areas that get a lot of snow in the winter time, but then it has Florida which never gets snow because it is too far south towards the equator. Then you have the Caribbean Islands, which contains a few countries and dozens of territories. There are over 7000 islands in the Caribbean, which many are not even inhabited. But the habitable island countries with people and hot climates are the Bahamas and Cuba. They rely mostly on tourism now for their economic growth. Cuba just recently settled diplomatic relations with the United States, so they are now accepting tourists to come to the country.

The landmass of North America is roughly 9.54 million square miles, which comes out to 16.5% of the total land area on Earth. The first humans to ever set foot in North America occurred during the Paleo-Indian era. This era started in 18,000 B.C. and ended around 8,000 B.C., so it was about 10,000 years ago. The present day people of North America were mostly influenced by the European settlers who arrived at the continent 500 years ago. These settlers were the first humans to really develop a more advanced civilization in North America that had economic prosperity and technological innovation. As of today, the United States and Canada are the most developed and wealthiest nations of the continent. Mexico is considered to be a newly industrialized nation, even though it was known for being a poor nation for a long time. But with its increase in globalization and international trade with the United States and other foreign countries, Mexico has become more developed as a result.

The author in Edmonton, Canada

The three main languages spoken in North America are English, French, and Spanish. The United States has mostly English and Spanish speaking people while Canada has mostly French and English speaking people. All of the countries in Central America are predominately Spanish speaking countries. This is due to the Spanish colonies that originated there. Before these colonies existed, a civilization known as "Maya" existed throughout Honduras, Guatemala, Belize and Mexico. The Mayans are the first civilization indigenous to the Americas that developed their own mathematical, astronomical, architecture and art systems. Many of their architectural structures still stand today, including the El Castillo pyramid in Mexico.

One country in North America that often gets overlooked is Greenland. This is actually a northern island country that is to the east of northern Canada. The Arctic Ocean and Atlantic Ocean surround Greenland. Since the country is so close to the arctic region, it is a very cold country with lots of snow and ice. But it has still been deemed as part of the North American continent. However, the strange thing about Greenland is that it is culturally and even politically part of Europe because it falls within the Kingdom of Denmark. This doesn't mean it is part of the European continent, but rather just part of the political and cultural rule of Europe. Geographically, it is still part of the North American continent because it is so close to the other North American countries. This may seem confusing when you look at Iceland, which is only 600 miles southeast of Greenland. Iceland is a much smaller island country, and it is considered part of the European continent and not the North American continent. Again, it is all about the convention of the boundaries. When you look at Iceland on a map it does appear a little bit closer to Europe than the mainland of North America. But still, it is closer to Greenland which is part of North America and yet it is considered to be a European country.

Chapter 5: South America

The continent of South America is located in the southern and western hemispheres of the world. There is also a small portion of the continent that falls within the northern hemisphere, but it is primarily considered a continent in the southern hemisphere. Scientists speculate that South America used to be adjoined with Africa millions of years ago, but the continents broke apart and slowly drifted away from each other until they got to the position in the Earth that they are at now. If you look at a map of the world and compare the shape of South America to that of Africa, you can see that South America looks like it could fit into the western side of Africa perfectly. This could be the result of the continents breaking apart and drifting away from each other millions of years ago.

before   **after**

The Pacific Ocean borders South America on the west and the Atlantic Ocean borders it on the east. Directly north of South America is North America, which is where Mexico, the United States and Canada are located. South America has twelve countries, or sovereign states, that make up the continent. These countries include Bolivia, Argentina, Chile, Brazil, Ecuador, Columbia, Peru, Paraguay, Guyana, Uruguay, Venezuela, and Suriname. There are also two states with no sovereignty; French Guiana and the Falkland Islands. French Guiana is actually a region of France which has a French overseas department there. The Falkland Islands is a territory ruled by the British, although Argentina has disputed this for quite some time. The entire South Amcrican continent is about 17.84 million kilometer squares.

The current total population of South America is about 388 million people. The continent ranks only fifth in having the most people, with Europe, North America, Africa and Asia having more people on its continents. The most populated nations in South America include Brazil, Argentina, Peru, Venezuela, and Colombia. The majority of the population in South America lives in the coastal towns and cities. As you get towards the middle of the continent you have less people living there. There is a good reason why this is the case. The Amazon rainforest takes up a good portion of the central northern region of the continent. This is a region with thousands of acres of rainforest that contains numerous swamps and wildlife. There are a few native tribes that live in this region, but it is certainly not a place for modernized people to live. There is no access to clean water or suitable living conditions. That is why people live near the coast because they have access to the ocean and more suitable climates to sustain proper living.

The South American Continent

The current ethnic and cultural influence over South America can primarily be attributed to the European settlers and conquerors that came to the continent hundreds of years ago. The Europeans brought with them immigrants and African slaves. These two groups associated themselves with the indigenous people of South America and eventually changed the ethnicity and culture of the region. Instead of it being primarily Hispanic people with one particular culture, it became a multicultural region with many different races and languages. The majority of South Americans speak Spanish and Portuguese. They are also more accustomed to Western traditions because of all the cultural influences brought to the continent by Europeans. This might sound confusing to say "Western traditions" since South America has always been in the Western hemisphere of the world. However, Western traditions are classified as the traditions brought to the west by Europeans. Some people call them European traditions, but it means the same thing.

South America does have many third world countries, which are countries that have a big economic gap between the number of rich people and the number of poor people. The gap is so big that there is hardly any middle class within these countries. The richest 10% of South Americans have more income than the bottom for 40% in countries like Chile, Paraguay, Colombia, Brazil and Bolivia. If you ever travel to cities within these countries then you will see numerous people living out on the streets in makeshift shacks and poor neighborhoods that are surprisingly right next to tall skyscrapers and luxury apartment buildings. The richest country in South America is Brazil, which thrives on manufacturing and exporting merchandise to other countries around the world. This is a $251 billion industry for the country. The next best export countries are Venezuela, Argentina and Chile. Venezuela brings in $93 billion, Argentina brings in $84 billion and Chile brings in $86 billion. Chances are if you want to survive living in South America then you should reside in one of these countries, so you can be close to the manufacturing jobs.

South America also has a tourism industry because of the beautiful landscapes and historical relics and architecture that the countries there have to offer. Millions of people travel to South America every year just to see them. Now in the 21st century, North American values and traditions are starting to influence more of these South American countries.

Chapter 6: Europe

Map of Europe

Europe is a continent in the North Hemisphere of the Earth and the westernmost part of the continental mass known as Eurasia, which is a combination of Europe and Asia. People often get confused trying to locate the boundary lines of Europe on a map because it is interconnected so much with Asia on its eastern side. However, the geological features that separate the two continents of Eurasia are the watershed divides of the Caucasus Mountains and Ural Mountains. You also have the Black Sea, Caspian Sea, the Turkish Straights waterways and the Ural River. These serve as the typical boundary lines of the continents when you look at them on a map. However, there are also political and cultural elements that greatly influence these boundary lines as well. To the north of Europe is the Arctic Ocean and to the west of Europe is the Atlantic Ocean. The Atlantic Ocean is the ocean that separates Europe from the western portion of the world that contains North America and South America. This is clearly visible on a map whereas the Europe and Asia divide has to be studied more closely on a map. As for the European countries, you have probably heard of many of them. Some of the most popular ones are Germany, France, Italy, England, Austria, Portugal, Greece, Poland, Romania, Spain, Ireland and many others. The majority of these countries have their own languages and cultural systems, even though they are geographically close to each other. But they still share western values in terms of technology and innovation.

The Houses of United Kingdom Parliament in London

Since Europe lies within the North Hemisphere, the climate on the continent is similar to that of North America. They both experience huge seasonal differences in terms of weather and temperature. Europe has warm summers brought on by Atlantic currents and harsh winters filled with snowy weather conditions. A great part of Asia also experiences these kinds of weather conditions because it lies primarily in the North Hemisphere as well. But those countries in Europe that are further from the Atlantic will experience more seasonal differences. You will find European countries are different in terms of culture as you go from east to west on the continent. Despite Europe having 50 countries and being the third most populated continent in the entire world, it is actually the second smallest continent in the world in terms of landmass. The total surface area of Europe is only 3.9 million square miles. In fact, the entire country of Russia is bigger than the entire continent of Europe and also takes up 40% of its landmass. It is estimated there are 740 million people living in Europe, which is approximately 11% of the global population. However, Europe is still a major economic power in the world and some countries of Europe have very advanced economies.

The author on Mount Teidi in Tenerife in the Canary Islands (Atlantic Ocean). The island is geographically on the African continent but politically and culturally belongs to Spain (Europe)

Greece is the one important country in Europe responsible for "Western culture." Remember that for a long time Europe was considered "Western" because it lies west of Asia and colonization of North America had not taken place yet. This is why Western culture in North America is actually European culture. Europe has had a long brutal history, but it is still responsible for the way of life that most civilized countries experience today. The Roman Empire that existed 2000 years ago really contributed to the migration of cultures and values throughout the continent. Europe became the first continent to lead the world in international affairs because of the Romans. By the time the 16th century came along, European nations began conquering other nations and inflicting their rule upon them. This started in the Americas, and then eventually led to Africa and most of Asia. All of the westernized cultures in these continents link back to the Europeans and their rule over these landmasses. You can find European territories in most continents in the world. Europe is associated with the British Virgin Islands in North America, French Guinea in South America and for awhile they ruled Hong Kong in Asia. With the increase in equality and fairness between nations, Europe is cutting down on its territories around the world that were originally established by force. In fact, the British even ruled the United States at one point before Americans decided to seek independence from Great Britain. This is why they had a Revolutionary War in which Great Britain lost, so they gave up the country to the Americans and it has been that way ever since. As for Europe, it is still a continent that is a thriving economic power in this world. Many of the countries in Europe are allies of the United States and other North American countries. It took hundreds of years for the nations of these continents to finally make peace with each other and now they finally have.

Chapter 7: Australia

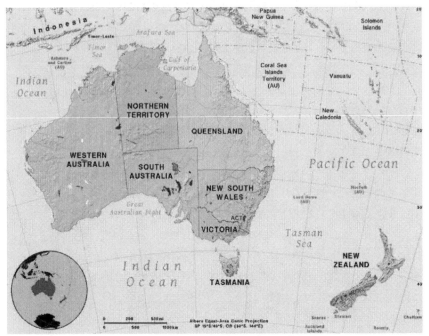
Austria and New Zealand

Australia is a continent in the southern hemisphere of the Earth. This continent is basically one big island that is not attached to any other landmasses like the other continents are. The only other continent that shares this similarity to Australia is Antarctica. However, the big difference between Australia and Antarctica is the fact that Australia is habitable for life. Australia contains a tropical climate that gives it warm enough weather to sustain life. The Australian continent also goes by more technical names, such as Meganesia, Australinea and Sahul. These names are used when distinguishing the continent of Australia from the country of Australia. Many people in other continents around the world assume that the Australian continent doesn't have any other countries in it. The truth is the Australian continent also contains Tasmania, Seram, New Guinea, Timor, and many surrounding islands around the mainland.

The Sydney Opera House theatre and concert hall complex in Sydney, Australia

Australia is the smallest continent out of all seven continents in the world. Australia lies on what is known as the continental shelf, which is an underwater landmass that exists in shallow water. This shallow water goes by the nickname "shelf sea." The reason for its shallowness has to do with the shelves being exposed during the interglacial and glacial periods. Before the ice age, the sea levels were lower and these shallow areas were actually land without water. But then about 20,000 years ago the sea levels rose and caused these shelves to form. There are also other shallow seas and water bodies that cover this landmass, including Torres Strait and the Arafura Sea, which lie between the country of Australia and New Guinea. There is also Bass Strait which lies between Tasmania and the Australian country. Sea levels have continued to rise over the years, which created separated islands away from mainland Australia. These two mountainous island nations are Tasmania and New Guinea. Believe it or not, these two countries are technically attached to mainland Australia. The only thing is the shallow seas between the nations make them look like separate lands.

New Zealand is an island country close to Australia. But despite its close distance on a global scale, New Zealand is not actually part of the Australian continent. In fact, it is one of the few nations that are not part of any continent. It used to be part of a continent named Zealandia, but it sank into the ocean after it broke apart from Australia about 23 million years ago. As for Australia itself, it has a total landmass of 3.3 million square miles. It is also the lowest continent on the Earth to sustain human life. People call it an island continent because the countries that make up the Australia are mostly islands. But again, the boundary lines of the continent were determined by human beings. It may seem confusing because it is a bunch of islands, but they have been deemed to be on the same continent because all of the landmass was attached to a bigger continent millions of years ago.

The nations of Australia are quite diverse because of the European settlers who came there hundreds of years ago. Today the nations are even more diverse because of all the immigrants who come from other countries to live in Australia. The reason they go there is because of the amazing weather conditions. Many parts of Australia have changing seasons like North America and some parts do not. This is very similar to the United States in the way certain states have snowy conditions and others don't. As of right now, Australia has roughly 36 million people living there out of all the nations on the continent. This might not seem like a lot of people, but you have to remember it is a small continent with small nations. The country of Australia is the biggest nation of the continent and it holds 23 million people. When you compare that to a country like the United States with 300 million people, then you can see how Australia is puny in comparison.

Australia surprisingly has a strong economy and conducts international trade all the time. It is obviously big in the tourism industry because of all the island resorts on the continent. But their agricultural industry is also huge because a lot of active farming is also going on. Australia is also a continent that doesn't have any dominate religious base. The area is so multicultural that there is no dominating religions or cultures in any of the countries. This may be another reason why people like to go there because they get a sense of freedom to be who they want to be without having to conform to any particular custom. If you ever get a chance to travel then you should choose Australia because there are so many exotic things to see and experience there.

Chapter 8: Antarctica

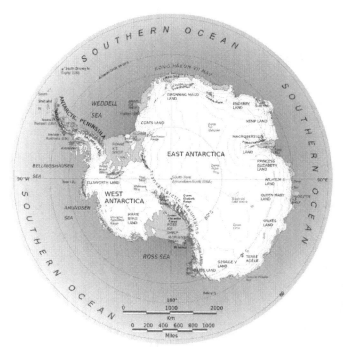

Map of Antarctica

Antarctica is a continent that lies within the southern hemisphere of the Earth. Actually, it lies on the very bottom of the Earth where the South Pole is located. This makes it the world's southernmost continent because if it were any further south it would actually be higher up north on the other end of the globe. Since Antarctica is located around the South Pole, it does not get a lot of direct sunlight. This is why it has such cold temperatures throughout the entire year. There are no climate differences between the seasons like you see in North America, Asia and Europe. No matter what time of year you go to visit Antarctica, it will always be cold and icy. In fact, the temperatures are colder than anything you would experience in North America or Europe.

When most people think about the continent known as "Antarctica" they think of a very cold place with not a lot of people. This would be an accurate assumption because there are not a lot of people on this continent. In fact, there are no people living on this continent because the freezing temperatures of Antarctica are just too severe to sustain human life. Now there are some humans that reside on the continent for research purposes only. These are typically scientists who represent a specific country. But you won't find any towns, cities or civilizations on the continent. The only life forms living in Antarctica are the wildlife.

One of the most famous animals living there are penguins. You will find plenty of penguins traveling together in groups. There are also foxes, mites, seals and many other creatures that can survive in cold temperatures. Other organisms include fungi, plants, algae and bacteria. If more people tried living on Antarctica then it would be a difficult process. There would have to be more buildings constructed with a lot of heaters which require a lot of power. Then the water and air of the continent would get polluted, which would then kill the wildlife and change the ecosystem of the environment altogether. Some scientists even say that tourism in Antarctica is polluting the water because of all the ships that pass through there. This ends up making the fish sick and kills the wildlife that eats the fish. If humans don't leave the continent alone then it might change all together.

Mount Murphy in Antarctica

The Southern Ocean (Antarctic Ocean) surrounds Antarctica, which makes it the only continent in the world to have a single ocean surrounding it completely. Antarctica is about 5.4 million square miles and is the fifth largest continent in the world. The two smaller continents are Europe and Australia. Antarctica is actually twice the size of Australia as far as landmass goes. Unfortunately, ice covers 98% of Antarctica which is another reason why it is not habitable. Now this isn't your everyday ice that you might find in North America during the wintertime. The ice in Antarctica is about 1.2 miles thick and extends everywhere except the Antarctic Peninsula in the northern region of the continent. Antarctica is not only the coldest and driest continent in the world, it is also the windiest. If that cold wind were to hit your naked skin then it would actually hurt you. That's how tough the cold wind is there.

Antarctica is technically a desert because of its barren landscape and uninhabitable conditions. You might think of deserts as only being desolate sandy landscapes that are hot and dry, but the word actually means any landscape that is desolate and unsuitable for life. Deserts also have very little precipitation, which means it doesn't snow or rain a whole lot in Antarctica. To give you an idea of just how cold it gets on the continent, the coldest temperature recorded was −129°F. Most people complain when it is 35°F outside in the wintertime. Can you imagine what −129°F below must feel like? You wouldn't want to imagine it because you would instantly get frostbite if a temperature that cold were to hit your naked skin. Even with a lot of heavy clothes and jackets on, you still wouldn't last too long outside in temperatures that cold. This is a big factor in why people cannot live on the continent.

Antarctica has no single government base. It is governed under the Antarctic Treaty System, which is a treaty that was signed back in 1959 by twelve different countries from around the world. Some of the main countries include the United States, Japan, France, Australia, Russia and the United Kingdom. The treaty basically gives permission for scientists of different countries to go to Antarctica for research purposes only. No one is allowed to perform military activities or commercial mining. This means no nuclear bomb testing or nuclear waste disposal allowed. The treaty is meant to protect the land with environmentally friendly practices only. There are over 4,000 scientists from around the world conducting experiments in Antarctica. If you want to visit the continent as a tourist then you can find cruise ships that will travel there from northern Russia and northern Alaska.

Chapter 9: Conclusion

The author at the Sultan Ahmed Mosque in Istanbul, Turkey. Istanbul is the only city that lies in two continents Asia and Europe

At this point, you should be very knowledgeable about the seven continents in the world. You know now that all of the continents are different in terms of culture, religion, climate, history and so forth. You also know that the continents of the world are identified by convention and not just by any natural separation, like from ocean water. People have used geological features to help them determine the boundary lines of the continents that are close together, such as Europe and Asia. These boundary lines could be mountains, streams or any other geological feature that is easy to identify as a boundary. In modern times, most nations have borders between other nations but not necessarily between continents. But the nations of different continents that are next to each other will still have borders. You just won't know it's a different continent unless you know which geological features to look for. In this book you have already learned about some of these geological features. Now you can impress your friends by being able to identify them because chances are your friends won't know what they are.

The continents discussed in this book are Africa, Asia, North America, Antarctica, South America, Australia and Europe. No matter where you are in the world, you come from one of these continents and are currently living on one of these continents. Some people stay in their native continent while others travel to new continents that they have never been before. It is important that everyone experiences these continents for themselves. But first, you have to understand their geography, cultural history and the environment they exist in. The people that reside on these continents are there for many different reasons. While some indigenous generations of families stayed on one continent for centuries, there was a lot of immigration that took place over the centuries as well. This allowed the continents we know today to become multicultural with a lot of diversity. So unless you understand the history of the continent you visit, you might be confused as to why certain cultures or religions currently exist there.

Hopefully now you can have a better appreciation for the way land on Earth is structured. If there were no continents then no one would feel like they belonged anywhere or to any particular government. Societies have outlined continents in order to understand the world better and for people to appreciate the region of the world they come from. If you are born in China then you can proudly call yourself Asian or Chinese. If you come from North America then you can simply call yourself an American because it is a region made up of many different races and cultures. What do you call yourself? Which region of the world do you come from? Maybe now you can answer these questions with a little more knowledge in your head about the continent you came from.

Thank you for reading and be sure to let everyone know what you have learned from this book!

References and further reading

National Geographic Little Kids First Big Book of the World (National Geographic Little Kids First Big Books), 2015. Elizabeth Carney. National Geographic Children's Books.

National Geographic Student World Atlas Fourth Edition, 2014 by National Geographic.

The Complete Book of Maps and Geography, Grades 3 – 6, 2009 by American Education Publishing.

Children Just Like Me: A Unique Celebration of Children Around the World, 1995. by Anabel Kindersley, Barnabas Kindersley et al.

World Civilizations and Cultures, Grades 5 - 8 (World History), 2012. Don Blattner. Mark Twain Media; Act Csm edition.

Explore Earth's Five Oceans (Explore the Continents), 2010. Bobbie Kalman. Crabtree Pub Co.

India For Kids: Amazing Facts About India, 2013. Shalu Sharma. CreateSpace Independent Publishing Platform.

Origins: The Evolution of Continents, Oceans and Life. 2001. Ron Redfern. University of Oklahoma Press.

Links to these books can be found on
http://www.nscientific.com/continents-of-the-world-for-kids/

Image credits:

Image in chapter 1 Introduction (globe); © [denisismagilov] / Dollar Photo Club

Image in chapter 3, Africa (Map of Africa); © [bogdanserban] / Dollar Photo Club

Image in chapter 5, South America (Planet earth before and after); © [robin2b] / Dollar Photo Club

53

Made in United States
Troutdale, OR
03/06/2024